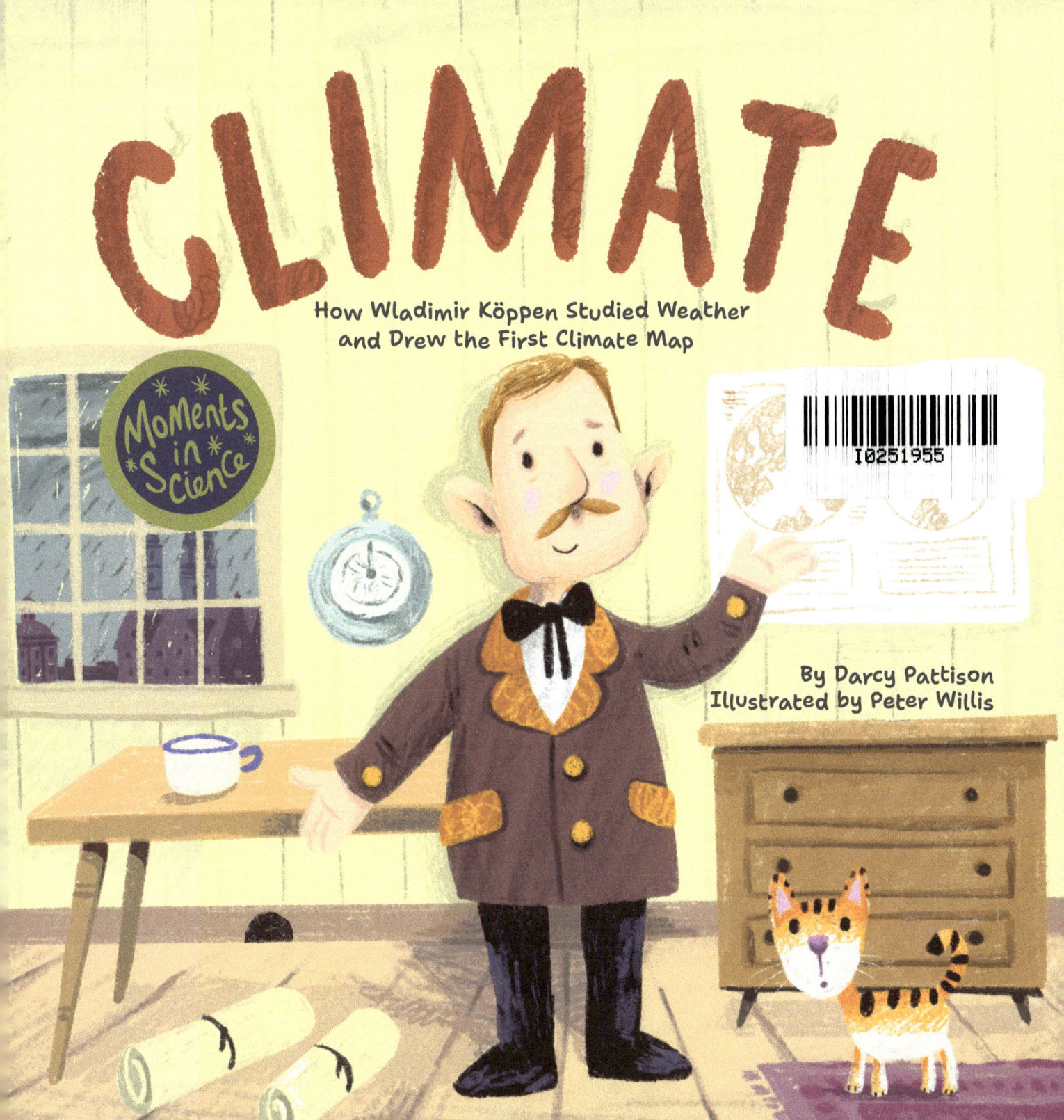

CLIMATE: How Wladimir Köppen Studied Weather and Drew the First Climate Map

Text copyright © 2025 Darcy Pattison
Illustrations copyright © 2025 Mims House

Mims House
1309 Broadway
Little Rock, AR 72202

MimsHouseBooks.com

Publisher's Cataloging-in-Publication Data

Names: Pattison, Darcy, author. | Willis, Peter, illustrator.
Title: Climate : how Wladimir Köppen studied weather and drew the first climate map / by Darcy Pattison; illustrations by Peter Willis.
Description: Little Rock, AR: Mims House, 2025. | Summary: In the late 1800s, weather was poorly understood. Wladimir Köppen, a German weatherman, gathered data from around the world to develop the world's first climate map.
Identifiers: LCCN: 2025900433 | ISBN: 9781629443058 (hardcover) | 9781629443065 (paperback) | 9781629443072 (ebook) | 9781629443089 (audio)
Subjects: LCSH Köppen, Wladimir Peter, 1846-1940. | Meteorologists--Biography--Juvenile literature. | Climatologists--Biography--Juvenile literature. | Climatology--Charts, diagrams, etc. | Meteorology--History--Juvenile literature. | Climatology--History--Juvenile literature. | BISAC JUVENILE NONFICTION / Science & Nature / Earth Sciences / Weather | JUVENILE NONFICTION / Biography & Autobiography / Science & Technology
Classification: LCC QC858.A2 .P 2025 | DDC 551.5/092--dc23

1854,

a train hissed and chugged, speeding out of St. Petersburg, Russia.

The Köppen family was aboard, traveling to their family estate 1,500 miles (2,414 km) to the south. 13-year-old Wladimir Köppen (VLAD-uh-meer KO-pun) leaned on the cool window, curious about the changing landscape of trees and grasses.

St. Petersburg, on the Baltic Sea, had rain and cool weather, which meant pine trees and swamps. But Moscow, which was far away from the sea, grew a mixed forest with birch, willow, and larch trees. Surrounded by the Black Sea, Crimea was warmer and grew oaks, pines, and firs. The family estate, Karabakh, meaning the Black Vineyard, was on the coast of the Black Sea, where the Köppens grew grape to make wine.

St. Petersburg

Moscow

Wladimir couldn't get the jumble of trees and plants out of his head.

Why did plants grow only in certain places? It had something to do with weather and climate, but how did it work? Someday he planned to understand it all.

After college, Wladimir took a job as the head weatherman in Hamburg, Germany, at the Deutsche Seewarte, the German Maritime Observatory on the Baltic Sea. At the time, people just knew the weather of their own town. But Wladimir's job was to create a daily weather map of Europe. It would show all the weather of the continent at the same time. It was a huge job, but Wladimir knew how to get it done.

His motto was... "Without hurry,

At first,

Wladimir helped set up weather stations along the coast. He sailed the North Sea—at the mercy of the weather—shivering in the cold winds. At each weather station, he checked the new weather instruments: thermometers to measure temperature, anemometers to measure wind speed, and rain gauges to measure precipitation.

He made sure the staff knew how to use the equipment.

Back at SeeWarte,

he received daily telegraphs with weather data from the stations. Each day—without hurry, without rest—Wladimir hunched over a map, putting in data until he completed the daily weather map. Wladimir even started adding weather predictions.

The collection of weather data gave Wladimir information over longer periods of time. A night owl, he didn't work well in the mornings, so he often stayed late at the office.

By candlelight, he spread papers across his desk and analyzed long-term trends. What could he learn about the weather by looking at data for a whole month, or a whole year? Or even for a decade?

Weather patterns started to become clear.

Across the world, other countries also set up weather stations and created daily weather maps. Now Wladimir studied bigger maps showing daily weather in North America and Europe.

But the Atlantic Ocean was blank— no data.

Wladimir started to fill in data on the Atlantic Ocean—without hurry, without rest.

Because Seewarte was a naval station, Wladimir could read logbooks with weather information from different ships that traveled the Atlantic Ocean. He installed weather instruments on the top of ship masts. The ships couldn't telegraph the data each day. But when they came back to port, Wladimir hurried to the harbor to talk to the sailors and gather the data.

Later, by candlelight, he read long into the night. Where had the data been recorded? What information had been captured? Wladimir struggled to make sense of the tangle of information.

Wladimir
Köppen
Atlas of the
Atlantic ocean
KÖPPEN
KÖPPEN
AUGUST
WLADIMIR
KÖPPEN

Slowly,

the ship's weather data started to make sense, and Wladimir carefully drew monthly wind maps of the Atlantic Ocean.

In 1883, he published *Atlas of the Atlantic Ocean*. Later, he added *Atlas of the Indian Ocean* (1891) and *Atlas of the Pacific Ocean* (1896). In 1899, he published his first book about climate entitled, *Climate Science*.

Now Wladimir

started looking skyward. Seewarte's daily weather data only showed temperature and precipitation on the surface of Earth. But the air went up and up. How far? What was the temperature up there? How much did it rain up there?

One blustery day, Wladimir walked across a Hamburg city park called Holy Ghost Field (Heiligengeistfeld). The sky swirled with the bright colors of children's kites dancing in the wind. What if he sent weather instruments into the atmosphere on kites?

In 1903,

Wladimir helped create a kite station to understand the great laboratory in the sky. The kites were anchored by a steel wire that fed from an electric winch. The winch was inside a small house that rotated so it could turn the kites into the wind. When a kite launched, the wire pulled freely, letting the kite soar. Often Wladimir added extra kites to the wire, creating a string of color snaking away into the wind.

Now, Wladimir went back to his original question:

Why do plants only grow in certain areas?

Without haste, without rest, Wladimir studied data from across the globe: daily, monthly, and yearly weather maps.

He searched for answers across land and sea, from sea level to the upper atmosphere.

To explain his ideas, he needed to draw a different kind of map—a world map.

Wladimir put together plant and weather information to divide the world's lands into areas with similar temperature and precipitation. Each plant species can only grow within a certain climate, or certain temperature ranges and precipitation levels. Wladimir's new map showed five climates across the globe: tropical, arid, temperate, continental, and polar.

It was the world's first climate map, a map that still helps us understand our planet Earth.

Wladimir Köppen, September 25, 1846 – June 22, 1940

Born in Russia, Köppen lived in St. Petersburg until his father retired in 1859, when the family moved to Crimea. The family's vineyard there was called Karabakh, which means Black Vineyard, because it was on the Black Sea.

At an early age, Köppen became passionately interested in scientific study. His greatest desire was to produce scientific work that would prove the truth about a question. He would study and study until he had the answer. He rarely gave up.

For college, Köppen studied botany and meterology in Russia and Germany before receiving his doctorate in 1870. He wrote, "My interests now revolve around the joy of knowledge, that is the anatomy of trematode larvae and frog throats, buzzard stomachs or the hearts of seagulls."

After he completed his education, he worked at Deutsche Seewarte, the German Maritime Observatory in Hamburg, Germany, for more than forty years. At first, he created daily weather maps, and then he began to forecast weather.

Köppen became interested in the upper atmosphere and how it affected weather. He was among the first to use weather balloons and kites to measure temperature, precipitation, and wind speed in the upper atmosphere. The balloon and kite work helped create a new field of study, aerology. Later, he published a cloud atlas. He invented machinery to launch weather kites and designed a station for the work. His rotating launch house turned in the wind to launch kites in the correct direction to attain the highest altitudes.

Köppens climate map was first published in 1884, and by 1900 he had refined his climate classification system. It was revised in 1918, and a final version was published in 1936 in his book, *The Geographic System of Climates*. German climatologist Rudolf Geiger (1894–1981) refined the system even more, with revisions in 1954 and 1961, so it was renamed the Köppen-Geiger climate classification system.

Besides being a scientist, Köppen was a scientific writer. He published over 560 journal articles and several books about his research. His life motto was "Without hurry, without rest."

THE KÖPPEN-GEIGER CLIMATE MAPS

This map shows the climate according to the Köppen-Geiger climate classification system, using data from 1901 to 1930. The Köppen-Geiger system defines five major categories of climate—tropical, arid, temperate, continental, and polar—and 30 sub-climates.

Tropical – Constant high temperature, with high annual precipitation (such as a rainforest) or dry winters (such as a savannah). Examples: Hilo, Hawaii; Miami, Florida; and Mumbai, India.

Dry – Arid desert, or semi-arid areas or steppes, with low precipitation.
Examples: Las Vegas, Nevada; Phoenix, Arizona; and the Sahara Desert in northern Africa.

THE KÖPPEN-GEIGER CLIMATE MAP
based on 1901–1930 data

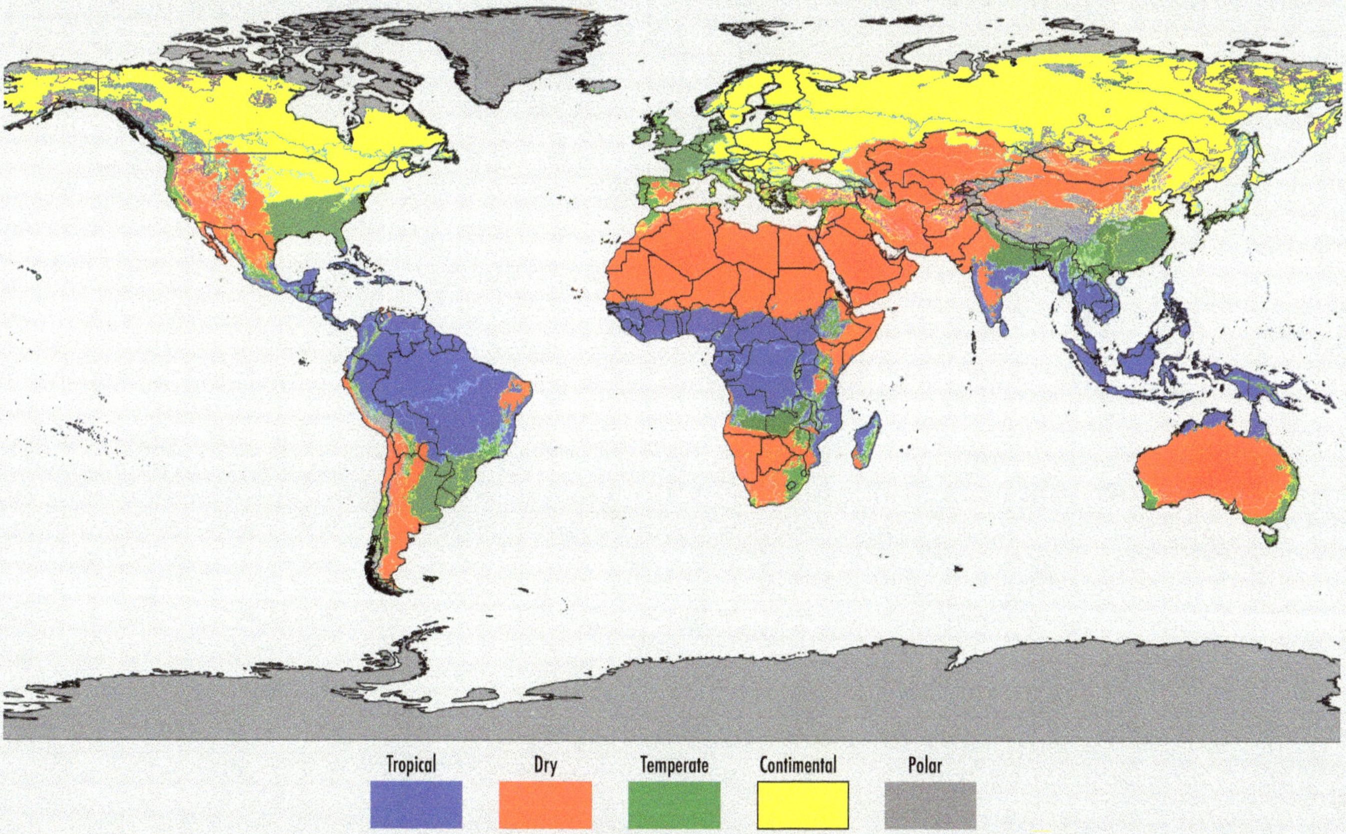

| Tropical | Dry | Temperate | Continental | Polar |

Source: Beck et al. (2023): High-resolution (1 km) Köppen-Geiger maps for 1901–2099 based on constrained CMIP6 projections, Scientific Data 10:724, doi:10.1038/s41597-023-02549-6.

Temperate – Moderate weather and moderate precipitations.
Examples: Los Angeles, California; Seattle, Washington; and Paris, France.

Continental – Interior of continents with moderate weather.
Examples: Chicago, Illinois; Salt Lake City, Utah; Fairbanks, Alaska; and Beijing, China.

Polar – The cold areas at the North and South Poles, or extremely high-altitude areas where it is constantly cold.
Examples: Denali, Alaska; Andes Mountains in Peru; and Antarctica.

WEATHER V. CLIMATE

Weather is the current conditions of temperature, wind, and precipitation in a certain location. For example, the daily weather of North America is different from the weather in Antarctica. Climate is a look at weather over a longer period of time. One hot day can't tell you much about climate, because climate depends on a certain location's average weather over months and years.

WEATHER INSTRUMENTS AVAILABLE TO KÖPPEN

Anemometer – Measures wind speed in miles per hour or kilometers per hour.
Barometer – Measures atmospheric pressure in atmospheres. One atmosphere (atm) is the average air pressure at sea level at a temperature of 59 degrees Fahrenheit (15 degrees Celsius).
Hygrometer – Measures humidity, or the percentage of water vapor in the air.
Rain gauge – Measures liquid precipitation in inches or centimeters.
Thermometer – Measures air or sea surface temperature in centigrade or Fahrenheit degrees.
Windsock or wind vane – Indicates wind direction. A windsock also indicates wind strength.

KÖPPEN'S WRITING PROCESS

1900 – While working on the classification of climates, Köppen wrote to his daughter:

"Some things I rewrite three or four times...[and the ideas] becomes clearer and clearer...[and I am] able to explain them to others."

SOURCES

Wegener-Köppen, Else. Ed. Wladimir Köoppen - Scholar for Life/Ein Gelehrtenleben für die Meteorologie. Original German edition and complete English translation with updated bibliography. Translated by Walter Obermiller. Borntraeger Science Publishers, Stuttgart 2018.
This is Köppen's autobiography, which was edited and completed after his death by his daughter, Else Wegener-Köppen. This version contains the complete German original text, along with a translated English version.

PHOTOGRAPHS

Photograph of Wladimir Köppen: Public domain.
Signature of Wladimir Köppen. Public domain.
Climate map: Beck, H.E., McVicar, T.R., Vergopolan, N., Alexis, B., Lutsko, N.J., Dufour, A., Zeng, Z., Jian, X., van Dijk, A.I.J.M., Miralles, D.G. "High-resolution (1 km) Köppen-Geiger maps for 1901–2099 based on constrained CMIP6 projections". Scientific Data. DOI:10.1038/s41597-023-02549-6.   https://commons.wikimedia.org/wiki/File:Koppen-Geiger_Map_v2_World_1901%E2%80%931930.svg
The map has been simplified to show only the five major categories of climate. In the same map sequence are more recent and predictive climate maps. If you are studying climate change, you may want to refer to these other maps.

www.ingramcontent.com/pod-product-compliance
Lightning Source LLC
Chambersburg PA
CBHW040225040426
42333CB00051B/3450

9781629443065